Lonnie Johnson

NASA Scientist and Inventor of the Super Soaker

by Lucia Raatma

PEBBLE
a capstone imprint

Little Explorer is published by Pebble, an imprint of Capstone.
1710 Roe Crest Drive,
North Mankato, Minnesota 56003
www.capstonepub.com

The name of the Smithsonian Institution and the sunburst logo are registered trademarks of the Smithsonian Institution. For more information, please visit www.si.edu.

Library of Congress Cataloging-in-Publication data is available on the Library of Congress website.
ISBN 978-1-9771-1413-6 (library binding)
ISBN 978-1-9771-1788-5 (paperback)
ISBN 978-1-9771-1417-4 (eBook PDF)
Summary: This book gives facts about Lonnie Johnson, a NASA scientist who invented the popular Super Soaker water toy.

Image Credits
Alabama Department of Archives and History: 10; Alamy: Jason Meyer, 13 (top), Keith Homan, 25 (top), Leon Werdinger, 4, Pete Jenkins, 25 (bottom); AP Photo: Atlanta Journal–Constitution/Jessica McGowan, 22; Getty Images: AFP/Paul J. Richards, 26, Mike McGregor, 18, The Enthusiast Network/Lynn Winelan, 6, The LIFE Images Collection/Thomas S. England, cover, 16, 27 (left); Library of Congress: 9; NASA: 14, JPL-Caltech, 15; Newscom: KRT/Akron Beacon Journal/Ken Love, 19 (top), KRT/Joe Rossi, 17; Photo Courtesy of the Johnson Research and Development Company Inc.: 5, 7, 8, 11, 12, 13 (bottom), 20, 21, 28; Shutterstock: denniz, 23, Kobby Dagan, 19 (bottom); U.S. Navy Photo by John F. Williams: 29; U.S. Patent and Trademark Office: 24, 27 (right)

Design Elements by Shutterstock

Editorial Credits
Editor: Michelle Parkin; Designer: Sarah Bennett; Media Researcher: Svetlana Zhurkin; Production Specialist: Tori Abraham

Our very special thanks to Emma Grahn, Spark!Lab Manager, Lemelson Center for the Study of Invention and Innovation, National Museum of American History. Capstone would also like to thank Kealy Gordon, Product Development Manager, and the following at Smithsonian Enterprises: Ellen Nanney, Licensing Manager; Brigid Ferraro, Vice President, Education and Consumer Products; and Carol LeBlanc, Senior Vice President, Education and Consumer Products.

All internet sites appearing in the back matter were available and accurate when this book was sent to press.

Printed in the United States of America.
PA99

TABLE OF CONTENTS

Bold words are in the glossary.

GET WET!

Whoosh! A rush of water streams from a Super Soaker. Playing with this water toy is a great way to stay cool on hot summer days.

A boy with the Super Soaker

Lonnie Johnson has **invented** many things. One of his most popular inventions is the Super Soaker water gun. Some of his inventions help solve complex problems. Some are fun!

Lonnie Johnson in front of the Super Soaker Wall at his company

The Super Soaker can spray water dozens of feet.

EARLY LIFE

Lonnie Johnson was born on October 6, 1949, in Mobile, Alabama. As a young boy, Lonnie watched his father fix things around the house. Lonnie liked to repair things too.

When he was 13, Lonnie attached a lawn mower engine to his Go-Kart racer. He raced it along a road until a police officer stopped him!

A Go-Kart from the 1950s

Lonnie's parents, David and Arline Johnson

Lonnie's father, David, worked as a driver at an Air Force base. His mother, Arline, was a nurse's aide. She also worked in a laundry business. Lonnie's parents picked cotton at a nearby farm to earn extra money.

When he was young, Lonnie liked to take his toys apart and put them back together. Lonnie took apart his sister's doll. He wanted to see how the doll's eyes opened and closed.

"As far back as I can remember, I was interested in devices and how they worked . . . "
—Lonnie Johnson

Lonnie around age 11

8

Lonnie's parents supported his interests. One day, Lonnie tried to make rocket fuel on top of the stove. It started a fire! But Lonnie's parents were not angry. Instead, they bought him a hot plate and told him to do his experiments outside.

One of Lonnie's role models was George Washington Carver, a black inventor.

George Washington Carver (1864–1943)

LIFE IN ALABAMA

Lonnie attended Williamson High School in Mobile. It was an all-black school. In the 1960s, many schools in Alabama were **segregated**. Black students couldn't go to the same schools as white students.

Alabama schools were segregated in the 1960s.

Lonnie's award-winning robot was named the Linex.

Lonnie and his robot

In 1968 Lonnie went to a science fair at the University of Alabama. He made a robot for the competition. Lonnie won first place.

Lonnie's friends called him The Professor.

Lonnie's love of inventions continued after high school. He received a **scholarship** to attend Tuskegee University in Alabama. There, he earned a bachelor's degree in mechanical engineering. Then he earned a master's degree in nuclear engineering.

Johnson as a student at Tuskegee University

Years after he graduated, Tuskegee University gave Johnson an honorary doctorate in science.

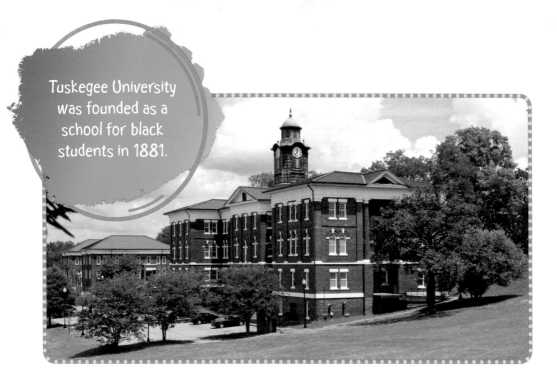

After college, Johnson worked as a research **engineer** at the U.S. Department of Energy's Oak Ridge National Laboratory. Then he joined the U.S. Air Force.

Johnson became a captain in the U.S. Air Force.

13

NASA SCIENTIST

In 1979 Johnson began working at NASA. He was a nuclear engineer in the Jet Propulsion Laboratory. He worked on many NASA space missions, including the spacecraft Galileo's mission to Jupiter. He also took part in the space **probe** Cassini's mission to Saturn.

NASA stands for the National Aeronautics and Space Administration.

Spacecraft Galileo reached Jupiter in 1995.

NASA's Jet Propulsion Laboratory in Pasadena, California

Johnson won many awards for his work at NASA.

Scientists inside the Jet Propulsion Lab

FUN INVENTION

Johnson worked at NASA during the day. In his free time, he continued to make his own inventions. One day in 1982, Johnson was at home working on a new invention. He was trying to make a better water pump.

Johnson made a high-powered **nozzle** for the pump and attached it to his bathroom sink. When he pressed the nozzle's lever, a strong stream of water shot across the bathroom! This gave Johnson the idea for the Super Soaker.

Johnson used a Super Soaker outside his home in 1992.

HOW A SUPER SOAKER WORKS

Water is stored in the reservoir. The pumping action builds up air **pressure**. When the trigger is pressed, it forces out a high-powered stream of water through the nozzle.

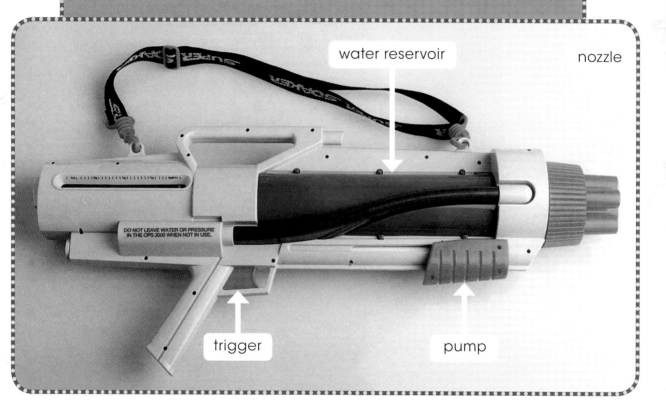

water reservoir

nozzle

trigger

pump

"I thought if I had a toy water gun
that was really high pressure and something
that a small kid could operate,
it would really be neat."

—Lonnie Johnson

Johnson called this toy the Power Drencher. Later, he changed the name to Super Soaker. He also improved the design. The first version of the Super Soaker had the water inside the water gun. Later, Johnson moved the water to a bottle on top of the toy.

Johnson with an early version of his Super Soaker invention in 2010

Johnson wanted to manufacture the water gun himself. But it was too expensive. In 1989 he sold the toy to the Larami Corporation. The company later became Hasbro. In 1991 the Super Soaker was the best-selling toy in the United States.

Hasbro is one of the largest toy companies in the world.

The Super Soaker become one of the most successful toys of all time. Over the years, the toy has earned $1 billion in sales. Johnson used the money he made from the Super Soaker to work on other projects. He wanted to invent items that helped the **environment**.

Johnson with one of his inventions

Johnson at the Johnson Research and Development Company

Johnson started the Johnson Research and Development Company in Atlanta, Georgia. The company makes rechargeable batteries. It produces engines and clean energy projects.

Coal, oil, and gas are called **fossil fuels**. Fossil fuels can harm the environment. And they are running out. Johnson knows it is important to use new sources of energy. Some of Johnson's inventions use **solar power**. Solar power comes from the sun. It can provide **electricity** to homes and other buildings.

Johnson at his lab in Atlanta in 2008

One of Johnson's inventions is called the Johnson Thermo-electrochemical Converter (JTEC). This can change heat from the sun into electricity. The JTEC does not cost much. Johnson hopes that it will be used in homes and power plants. One day, it could even power a spacecraft.

The magazine *Popular Mechanics* named the JTEC one of the top 10 new world-changing innovations in 2008.

SO MANY INVENTIONS

Johnson continues to invent new products. One invention curls and dries hair. He also created a portable multimedia projector. The projector has a DVD player, a channel tuner, and a remote-control receiver.

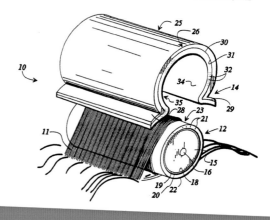

Johnson's patent for his hair-drying curler invention

A Super Soaker from 2016

Johnson continues to make toys. He has made many new versions of the Super Soaker. He has also invented foam darts for the Nerf brand at Hasbro.

Nerf stands for nonexpanding recreational foam.

LATER YEARS

Johnson is always trying new things. Some ideas work, and others don't. He has to use his imagination when inventions fail. He tries over and over until the product works.

Johnson applies for a **patent** for each one of his inventions. A patent is a document that proves the idea belongs to him. Getting a patent can take a long time. But it is important to keep others from taking credit and getting money for his inventions.

The U.S. Patent and Trademark Office in Virginia

Johnson with his patent for the Super Soaker water toy

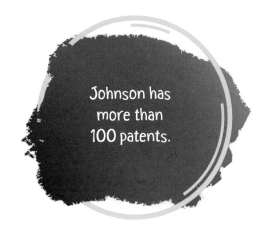

Johnson has more than 100 patents.

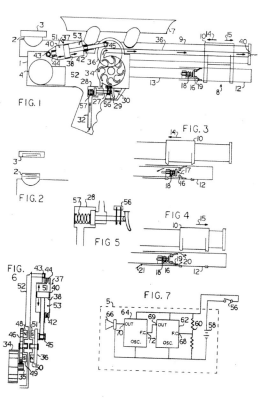

Johnson has four children. Today he lives in Georgia with his wife, Linda Moore. Johnson is active in his community. He is one of the leaders of the Georgia Alliance for Children. This group protects the rights of children in the state. Johnson is also a member of the 100 Black Men of Atlanta. This group supports young black students and helps pay for their education.

The mayor of Marietta, Georgia, declared February 25, 1994, Lonnie G. Johnson Day.

Johnson continues to speak about energy and environmental issues.

Johnson has been very successful. He has earned a lot of money. But Johnson wants to keep working. He wants to encourage young inventors. And he knows there are still more inventions to create.

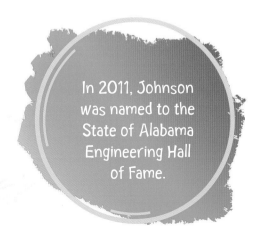

In 2011, Johnson was named to the State of Alabama Engineering Hall of Fame.

"I feel like I have been blessed with a talent or a skill, and what motivates me . . . is to try to have a positive impact on the environment and people's quality of life."

—Lonnie Johnson

GLOSSARY

electricity (i-lek-TRISS-uh-tee)—a natural force that can be used to make light and heat or make machines work

engineer (en-juh-NEER)—someone trained to design and build machines, vehicles, bridges, roads, and other structures

environment (in-VY-ruh-muhnt)—the natural world of the land, water, and air

fossil fuel (FAH-suhl FYOOL)—a natural fuel formed from the remains of plants and animals; coal, oil, and natural gas are fossil fuels

invent (in-VENT)—to think up and create something new

nozzle (NOZ-uhl)—a spout that directs the flow of liquid from the end of a hose or tube

patent (PAT-uhnt)—a legal document giving the inventor sole rights to make and sell an item he or she invented

pressure (PRESH-ur)—a force made by pressing on something

probe (PROHB)—a small, unmanned vehicle used to explore outer space

scholarship (SKOL-ur-ship)—money given to a student to pay for school

segregate (SEG-ruh-gate)—to keep people apart based on their skin color; segregation is illegal in the United States

solar power (SOH-lur POU-ur)—energy from the sun that can be used for heating and electricity

CRITICAL THINKING QUESTIONS

1. In your own words, describe how Lonnie Johnson invented the Super Soaker.

2. Why do you think patents are important?

3. Imagine you are using a Super Soaker water toy. Would it be more fun to get soaked or soak someone else?

READ MORE

Barton, Chris. *Whoosh!: Lonnie Johnson's Super-Soaking Stream of Inventions.* Watertown, MA: Charlesbridge, 2016

Meister, Cari. *Totally Amazing Facts About Outrageous Inventions.* North Mankato, MN: Capstone Press, 2017.

Schwartz, Heather E. *Super Soaker Inventor Lonnie Johnson.* Minneapolis: Lerner Publications, 2017.

INTERNET SITES

Famous Inventors—Lonnie Johnson
https://www.famousinventors.org/lonnie-johnson

Meet Lonnie Johnson, the Man Behind the Super Soaker
https://invention.si.edu/meet-lonnie-johnson-man-behind-super-soaker

INDEX